I0103612

THE MALAYAN PAPILIONIDÆ

OR SWALLOW-TAILED BUTTERFLIES, AS ILLUSTRATIVE OF THE THEORY OF NATURAL SELECTION
(1864/1870)

BY

ALFRED RUSSEL WALLACE

Copyright © 2013 Read Books Ltd.
This book is copyright and may not be
reproduced or copied in any way without
the express permission of the publisher in writing

British Library Cataloguing-in-Publication Data
A catalogue record for this book is available from the
British Library

Alfred Russel Wallace

Alfred Russel Wallace was born on 8th January 1823 in the village of Llanbadoc, in Monmouthshire, Wales.

At the age of five, Wallace's family moved to Hertford where he later enrolled at Hertford Grammar School. He was educated there until financial difficulties forced his family to withdraw him in 1836. He then boarded with his older brother John before becoming an apprentice to his eldest brother, William, a surveyor. He worked for William for six years until the business declined due to difficult economic conditions.

After a brief period of unemployment, he was hired as a master at the Collegiate School in Leicester to teach drawing, map-making, and surveying. During this time he met the entomologist Henry Bates who inspired Wallace to begin collecting insects. He and bates continued exchanging letters after Wallace left teaching to pursue his surveying career. They corresponded on prominent works of the time such as Charles Darwin's *The Voyage of the Beagle* (1839) and Robert Chamber's *Vestiges of the Natural History of Creation* (1844).

Wallace was inspired by the travelling naturalists of the day and decided to begin his exploration career collecting specimens in the Amazon rainforest. He explored the Rio Negra for four years, making notes on the peoples and languages he encountered as well as the geography, flora, and fauna. On his return voyage his ship, Helen, caught fire and he and the crew were stranded for ten days before being picked up by the Jordeson, a brig travelling from Cuba to London. All of his specimens aboard Helen had been lost.

After a brief stay in England he embarked on a journey to the Malay Archipelago (now Singapore, Malaysia, and Indonesia). During this eight year period he collected more than 126,000 specimens, several thousand of which represented new species to science. While travelling, Wallace refined his thoughts about evolution and in 1858 he outlined his theory of natural selection in an article he sent to Charles Darwin. This was published in the same year along with Darwin's own theory. Wallace eventually published an account of his travels *The Malay Archipelago* in 1869, and it became one of the most popular books of scientific exploration in the 19th century.

Upon his return to England, in 1862, Wallace became a staunch defender of Darwin's landmark work *On the Origin of Species* (1859). He wrote responses to those critical of the theory of natural selection, including 'Remarks on the Rev. S. Haughton's Paper on the Bee's Cell, And on the Origin of Species' (1863) and 'Creation by Law' (1867). The former of these was particularly pleasing to Darwin. Wallace also published important papers such as 'The Origin of Human Races and the Antiquity of Man Deduced from the Theory of 'Natural Selection" (1864) and books, including the much cited *Darwinism* (1889).

Wallace made a huge contribution to the natural sciences and he will continue to be remembered as one of the key figures in the development of evolutionary theory.

Wallace died on 7th November 1913 at the age of 90. He is buried in a small cemetery at Broadstone, Dorset, England.

THE MALAYAN PAPILIONIDÆ OR SWALLOW-TAILED BUTTERFLIES, AS ILLUSTRATIVE OF THE THEORY OF NATURAL SELECTION

(1864/1870)

Special Value of the Diurnal Lepidoptera for enquiries of this nature.

When the naturalist studies the habits, the structure, or the affinities of animals, it matters little to which group he especially devotes himself; all alike offer him endless materials for observation and research. But, for the purpose of investigating the phenomena of geographical distribution and of local, sexual, or general variation, the several groups differ greatly in their value and importance. Some have too limited a range, others are not sufficiently varied in specific forms, while, what is of most importance, many groups have not received that amount of attention over the whole region they inhabit, which could furnish materials sufficiently approaching to completeness to enable us to arrive at any accurate conclusions as to the phenomena they present as a whole. It is in those groups which are, and have long been, favourites with collectors, that the student of distribution and variation will find his materials the most satisfactory, from their comparative completeness.

Pre-eminent among such groups are the diurnal Lepidoptera or Butterflies, whose extreme beauty and endless diversity have led to their having been assiduously collected in all parts of the world, and to the numerous species and varieties having been figured in a series of magnificent works, from those of Cramer, the contemporary of Linnæus, down to the inimitable productions of our own Hewitson.[1] But, besides their abundance, their universal distribution, and the great attention that has been paid to them, these insects have other qualities that especially adapt them to elucidate the branches of inquiry already alluded to. These are, the immense development and peculiar structure of the wings, which not only vary in form more than those of any other insects, but offer on both surfaces an endless variety of pattern, colouring, and texture. The scales, with which they are more or less completely covered, imitate the rich hues and delicate surfaces of satin or of velvet, glitter with metallic lustre, or glow with the changeable tints of the opal. This delicately painted surface acts as a register of the minutest differences of organization--a shade of colour, an additional streak or spot, a slight modification of outline continually recurring with the greatest regularity and fixity, while the body and all its other members exhibit no appreciable change. The wings of Butterflies, as Mr. Bates has well put it, "serve as a tablet on which Nature writes the story of the modifications of species;" they enable us to perceive changes that would otherwise be uncertain and difficult of observation, and exhibit to us on an enlarged scale the effects of the climatal and other physical conditions which influence more or less profoundly the organization of every living thing.

A proof that this greater sensibility to modifying causes is not imaginary may, I think, be drawn from the consideration, that while the Lepidoptera as a whole are of all insects the least essentially varied in form, structure, or habits, yet in the number of their specific forms they are not much inferior to those orders which range over a much wider field of nature, and exhibit more deeply seated structural modifications. The Lepidoptera are all vegetable-feeders in their larva-state, and suckers of juices or other liquids in their perfect form. In their most widely separated groups they differ but little from a common type, and offer comparatively unimportant modifications of structure or of habits. The Coleoptera, the Diptera, or the Hymenoptera, on the other hand, present far greater and more essential variations. In either of these orders we have both vegetable and animal-feeders, aquatic, and terrestrial, and parasitic groups. Whole families are devoted to special departments in the economy of nature. Seeds, fruits, bones, carcases, excrement, bark, have each their special and dependent insect tribes from among them; whereas the Lepidoptera are, with but few exceptions, confined to the one function of devouring the foliage of living vegetation. We might therefore anticipate that their species-population would be only equal to that of sections of the other orders having a similar uniform mode of existence; and the fact that their numbers are at all comparable with those of entire orders, so much more varied in organization and habits, is, I think, a proof that they are in general highly susceptible of specific modification.

Question of the rank of the Papilionidæ.

The Papilionidæ are a family of diurnal Lepidoptera which have hitherto, by almost universal consent, held the first rank in the order; and though this position has recently been denied them, I cannot altogether acquiesce in the reasoning by which it has been proposed to degrade them to a lower rank. In Mr. Bates's most excellent paper on the Heliconidæ, (published in the Transactions of the Linnæan Society, vol. xxiii.,) he claims for that family the highest position, chiefly because of the imperfect structure of the fore legs, which is there carried to an extreme degree of abortion, and thus removes them further than any other family from the Hesperidæ and Heterocera, which all have perfect legs. Now it is a question whether any amount of difference which is exhibited merely in the imperfection or abortion of certain organs, can establish in the group exhibiting it a claim to a high grade of organization; still less can this be allowed when another group along with perfection of structure in the same organs, exhibits modifications peculiar to it, together with the possession of an organ which in the remainder of the order is altogether wanting. This is, however, the position of the Papilionidæ. The perfect insects possess two characters quite peculiar to them. Mr. Edward Doubleday, in his "Genera of Diurnal Lepidoptera," says, "The Papilionidæ may be known by the apparently four-branched median nervule and the spur on the anterior tibiæ, characters found in no other family." The four-branched median nervule is a character so constant, so peculiar, and so well marked, as to enable a person to tell, at a glance at the wings only of a butterfly, whether it does or does not belong to this family; and I am not aware that any other group of butterflies, at all comparable to this in extent and modifications of form, possesses a character in its neuration to which the same degree of certainty can be attached. The spur on the anterior tibiæ is also found in some of the Hesperidæ, and is therefore supposed to show a direct affinity between the two groups: but I do not imagine it can counterbalance the differences in neuration and in every other part of their organization.

The most characteristic feature of the Papilionidæ, however, and that on which I think insufficient stress has been laid, is undoubtedly the peculiar structure of the larvæ. These all possess an extraordinary organ situated on the neck, the well-known Y-shaped tentacle, which is entirely concealed in a state of repose, but which is capable of being suddenly thrown out by the insect when alarmed. When we consider this singular apparatus, which in some species is nearly half an inch long, the arrangement of muscles for its protrusion and retraction, its perfect concealment during repose, its blood-red colour, and the suddenness with which it can be thrown out, we must, I think, be led to the conclusion that it serves as a protection to the larva, by startling and frightening away some enemy when about to seize it, and is thus one of the causes which has led to the wide extension and maintained the permanence of this now dominant group. Those who believe that such peculiar structures can only have arisen by very minute successive variations, each one advantageous to its possessor, must see, in the possession of such an organ by one group, and its complete absence in every other, a proof of a very ancient origin and of very long-continued modification. And such a positive structural addition to the organization of the family, subserving an important function, seems to me alone sufficient to warrant us in considering the Papilionidæ as the most highly developed portion of the whole order, and thus in retaining it in the position which the size, strength, beauty, and general structure of the perfect insects have been generally thought to deserve.

In Mr. Trimen's paper on "Mimetic Analogies among African Butterflies," in the Transactions of the Linnæan Society, for 1868, he has argued strongly in favour of Mr. Bates' views as to the higher position of the Danaidæ and the lower grade of the Papilionidæ, and has adduced, among other facts, the undoubted resemblance of the pupa of Parnassius, a genus of Papilionidæ, to that of some Hesperidæ and moths. I admit, therefore, that he has proved the Papilionidæ to have retained several characters of the nocturnal Lepidoptera which the Danaidæ have lost, but I deny that they are therefore to be considered lower in the scale of organization. Other characters may be pointed out which indicate that they are farther removed from the moths even than the Danaidæ. The club of the antennæ is the most prominent and most constant feature by which butterflies may be distinguished from moths, and of all butterflies the Papilionidæ have the most beautiful and most perfectly developed clubbed antennæ. Again, butterflies and moths are broadly characterised by their diurnal and nocturnal habits respectively, and the Papilionidæ, with their close allies the Pieridæ, are the most pre-eminently diurnal of butterflies, most of them lovers of sunshine, and not presenting a single crepuscular species. The great group of the Nymphalidæ, on the other hand (in which Mr. Bates includes the Danaidæ and Heliconidæ as sub-families), contains an entire sub-family (Brassolidæ) and a number of genera, such as Thaumantis, Zeuxidia, Pavonia, &c., of crepuscular habits, while a large proportion of the Satyridæ and many of the Danaidæ are shade-loving butterflies. This question, of what is to be considered the highest type of any group of organisms, is one of such general interest to naturalists that it will be well to consider it a little further, by a comparison of the Lepidoptera with some groups of the higher animals.

Mr. Trimen's argument, that the lepidopterous type, like that of birds, being preeminently aërial, "therefore a diminution of the ambulatory organs, instead of being a sign of inferiority, may very possibly indicate a higher, because a more thoroughly aërial form," is certainly unsound, for it would imply that the most aërial of birds (the swift and the frigate-birds, for example) are the highest in the scale of bird-organization, and the more so on account of their feet being very ill adapted for walking. But no ornithologist has ever so classed them, and the claim to the highest rank among birds is only disputed between three groups, all very far removed from these. They are--1st. The Falcons, on account of their general perfection, their rapid flight, their piercing vision, their perfect feet armed with retractile claws, the beauty of their forms, and the ease and rapidity of their motions; 2nd. The Parrots, whose feet, though ill-fitted for walking, are perfect as prehensile organs, and which possess large brains with great intelligence, though but moderate powers of flight; and, 3rd. The Thrushes or Crows, as typical of the perching birds, on account of the well-balanced development of their whole structure, in which no organ or function has attained an undue prominence.

Turning now to the Mammalia, it might be argued that as they are pre-eminently the terrestrial type of vertebrates, to walk and run well is essential to the typical perfection of the group; but this would give the superiority to the horse, the deer, or the hunting leopard, instead of to the Quadrumana. We seem here to have quite a case in point, for one group of Quadrumana, the Lemurs, is undoubtedly nearer to the low Insectivora and Marsupials than the Carnivora or the Ungulata, as shown among other characters by the Opossums possessing a hand with perfect opposable thumb, closely resembling that of some of the Lemurs; and by the curious Galeopithecus, which is sometimes classed as a Lemur, and sometimes with the Insectivora. Again, the implacental mammals, including the Ornithodelphia and the Marsupials, are admitted to be lower than the placental series. But one of the distinguishing characters of the Marsupials is that the young are born blind and exceedingly imperfect, and it might therefore be argued that those orders in which the young are born most perfect are the highest, because farthest from the low Marsupial type. This would make the Ruminants and Ungulata higher than the Quadrumana or the Carnivora. But the Mammalia offer a still more remarkable illustration of the fallacy of this mode of reasoning, for if there is one character more than another which is essential and distinctive of the class, it is that from which it derives its name, the possession of mammary glands and the power of suckling the young. What more reasonable, apparently, than to argue that the group in which this important function is most developed, that in which the young are most dependent upon it, and for the longest period, must be the highest in the Mammalian scale of organization? Yet this group is the Marsupial, in which the young commence suckling in a fœtal condition, and continue to do so till they are fully developed, and are therefore for a long time absolutely dependent on this mode of nourishment.

These examples, I think, demonstrate that we cannot settle the rank of a group by a consideration of the degree in which certain characters resemble or differ from those in what is admitted to be a lower group; and they also show that the highest group of a class may be more closely connected to one of the lowest, than some other groups which have developed laterally and diverged farther from the parent type, but which yet, owing to want of balance or too great specialization in their structure, have never reached a high grade of organization. The Quadrumana afford a very valuable illustration, because, owing to their undoubted affinity with man, we feel certain that they are really higher than any other order of Mammalia, while at the same time they are more distinctly allied to the lowest groups than many others. The case of the Papilionidæ seems to me so exactly parallel to this, that, while I admit all the proofs of affinity with the undoubtedly lower groups of Hesperidæ and moths, I yet maintain that, owing to the complete and even development of every part of their organization, these insects best represent the highest perfection to which the butterfly type has attained, and deserve to be placed at its head in every system of classification.

Distribution of the Papilionidæ.

The Papilionidæ are pretty widely distributed over the earth, but are especially abundant in the tropics, where they attain their maximum of size and beauty, and the greatest variety of form and colouring. South America, North India, and the Malay Islands are the regions where these fine insects occur in the greatest profusion, and where they actually become a not unimportant feature in the scenery. In the Malay Islands in particular, the giant Ornithopteræ may be frequently seen about the borders of the cultivated and forest districts, their large size, stately flight, and gorgeous colouring rendering them even more conspicuous than the generality of birds. In the shady suburbs of the town of Malacca two large and handsome Papilios (Memnon and Nephelus) are not uncommon, flapping with irregular flight along the roadways, or, in the early morning, expanding their wings to the invigorating rays of the sun. In Amboyna and other towns of the Moluccas, the magnificent Deiphobus and Severus, and occasionally even the azure-winged Ulysses, frequent similar situations, fluttering about the orange-trees and flower-beds, or sometimes even straying into the narrow bazaars or covered markets of the city. In Java the golden-dusted Arjuna may often be seen at damp places on the roadside in the mountain districts, in company with Sarpedon, Bathycles, and Agamemnon, and less frequently the beautiful swallow-tailed Antiphates. In the more luxuriant parts of these islands one can hardly take a morning's walk in the neighbourhood of a town or village without seeing three or four species of Papilio, and often twice that number. No less than 130 species of the family are now known to inhabit the Archipelago, and of these ninety-six were collected by myself. Thirty species are found in Borneo, being the largest number in any one island, twenty-three species having been obtained by myself in the vicinity of Sarawak; Java has twenty-eight species; Celebes twenty-four, and the Peninsula of Malacca, twenty-six species. Further east the numbers decrease; Batchian producing seventeen, and New Guinea only fifteen, though this number is certainly too small, owing to our present imperfect knowledge of that great island.

Definition of the word Species.

In estimating these numbers I have had the usual difficulty to encounter, of determining what to consider species and what varieties. The Malayan region, consisting of a large number of islands of generally great antiquity, possesses, compared to its actual area, a great number of distinct forms, often indeed distinguished by very slight characters, but in most cases so constant in large series of specimens, and so easily separable from each other, that I know not on what principle we can refuse to give them the name and rank of species. One of the best and most orthodox definitions is that of Pritchard, the great ethnologist, who says, that "*separate origin and distinctness of race, evinced by a constant transmission of some characteristic peculiarity of organization*," constitutes a species. Now leaving out the question of "origin," which we cannot determine, and taking only the proof of separate origin, "*the constant transmission of some characteristic peculiarity of organization*," we have a definition which will compel us to neglect altogether the *amount* of difference between any two forms, and to consider only whether the differences that present themselves are *permanent.* The rule, therefore, I have endeavoured to adopt is, that when the difference between two forms inhabiting separate areas seems quite constant, when it can be defined in words, and when it is not confined to a single peculiarity only, I have considered such forms to be species. When, however, the individuals of each locality vary among themselves, so as to cause the distinctions between the two forms to become inconsiderable and indefinite, or where the differences, though constant, are confined to one particular only, such as size, tint, or a single point of difference in marking or in outline, I class one of the forms as a variety of the other.

I find as a general rule that the constancy of species is in an inverse ratio to their range. Those which are confined to one or two islands are generally very constant. When they extend to many islands, considerable variability appears; and when they have an extensive range over a large part of the Archipelago, the amount of unstable variation is very large. These facts are explicable on Mr. Darwin's principles. When a species exists over a wide area, it must have had, and probably still possesses, great powers of dispersion. Under the different conditions of existence in various portions of its area, different variations from the type would be selected, and, were they completely isolated, would soon become distinctly modified forms; but this process is checked by the dispersive powers of the whole species, which leads to the more or less frequent intermixture of the incipient varieties, which thus become irregular and unstable. Where, however, a species has a limited range, it indicates less active powers of dispersion, and the process of modification under changed conditions is less interfered with. The species will therefore exist under one or more permanent forms according as portions of it have been isolated at a more or less remote period.

Laws and Modes of Variation.

What is commonly called variation consists of several distinct phenomena which have been too often confounded. I shall proceed to consider these under the heads of--1st, simple variability; 2nd, polymorphism; 3rd, local forms; 4th, co-existing varieties; 5th, races or subspecies; and 6th, true species.

1. *Simple variability.*--Under this head I include all those cases in which the specific form is to some extent unstable. Throughout the whole range of the species, and even in the progeny of individuals, there occur continual and uncertain differences of form, analogous to that variability which is so characteristic of domestic breeds. It is impossible usefully

to define any of these forms, because there are indefinite gradations to each other form. Species which possess these characteristics have always a wide range, and are more frequently the inhabitants of continents than of islands, though such cases are always exceptional, it being far more common for specific forms to be fixed within very narrow limits of variation. The only good example of this kind of variability which occurs among the Malayan Papilionidæ is in Papilio Severus, a species inhabiting all the islands of the Moluccas and New Guinea, and exhibiting in each of them a greater amount of individual difference than often serves to distinguish well-marked species. Almost equally remarkable are the variations exhibited in most of the species of Ornithoptera, which I have found in some cases to extend even to the form of the wing and the arrangement of the nervures. Closely allied, however, to these variable species are others which, though differing slightly from them, are constant and confined to limited areas. After satisfying oneself, by the examination of numerous specimens captured in their native countries, that the one set of individuals are variable and the others are not, it becomes evident that by classing all alike as varieties of one species we shall be obscuring an important fact in nature; and that the only way to exhibit that fact in its true light is to treat the invariable local form as a distinct species, even though it does not offer better distinguishing characters than do the extreme forms of the variable species. Cases of this kind are the Ornithoptera Priamus, which is confined to the islands of Ceram and Amboyna, and is very constant in both sexes, while the allied species inhabiting New Guinea and the Papuan Islands is exceedingly variable; and in the island of Celebes is a species closely allied to the variable P. Severus, but which, being exceedingly constant, I have described as a distinct species under the name of Papilio Pertinax.

2. *Polymorphism or dimorphism.*--By this term I understand the co-existence in the same locality of two or more distinct forms, not connected by intermediate gradations, and all of which are occasionally produced from common parents. These distinct forms generally occur in the female sex only, and their offspring, instead of being hybrids, or like the two parents, appear to reproduce all the distinct forms in varying proportions. I believe it will be found that a considerable number of what have been classed as *varieties* are really cases of polymorphism. Albinoism and melanism are of this character, as well as most of those cases in which well-marked varieties occur in company with the parent species, but without any intermediate forms. If these distinct forms breed independently, and are never reproduced from a common parent, they must be considered as separate species, contact without intermixture being a good test of specific difference. On the other hand, intercrossing without producing an intermediate race is a test of dimorphism. I

consider, therefore, that under any circumstances the term "variety" is wrongly applied to such cases.

The Malayan Papilionidæ exhibit some very curious instances of polymorphism, some of which have been recorded as varieties, others as distinct species; and they all occur in the female sex. Papilio Memnon is one of the most striking, as it exhibits the mixture of simple variability, local and polymorphic forms, all hitherto classed under the common title of varieties. The polymorphism is strikingly exhibited by the females, one set of which resemble the males in form, with a variable paler colouring; the others have a large spatulate tail to the hinder wings and a distinct style of colouring, which causes them closely to resemble P. Coon, a species having the two sexes alike and inhabiting the same countries, but with which they have no direct affinity. The tailless females exhibit simple variability, scarcely two being found exactly alike even in the same locality. The males of the island of Borneo exhibit constant differences of the under surface, and may therefore be distinguished as a local form, while the continental specimens, as a whole, offer such large and constant differences from those of the islands, that I am inclined to separate them as a distinct species, to which the name P. Androgens (Cramer) may be applied. We have here, therefore, distinct species, local forms, polymorphism, and simple variability, which seem to me to be distinct phenomena, but which have been hitherto all classed together as varieties. I may mention that the fact of these distinct forms being one species is doubly proved. The males, the tailed and tailless females, have all been bred from a single group of the larvæ, by Messrs. Payen and Bocarmé, in Java, and I myself captured, in Sumatra, a male P. Memnon, and a tailed female P. Achates, under circumstances which led me to class them as the same species.

Papilio Pammon offers a somewhat similar case. The female was described by Linnæus as P. Polytes, and was considered to be a distinct species till Westermann bred the two from the same larvæ (see Boisduval, "Species Général des Lépidoptères,"). They were therefore classed as sexes of one species by Mr. Edward Doubleday, in his "Genera of Diurnal Lepidoptera," in 1846. Later, female specimens were received from India closely resembling the male insect, and this was held to overthrow the authority of M. Westermann's observation, and to re-establish P. Polytes as a distinct species; and as such it accordingly appears in the British Museum List of Papilionidæ in 1856, and in the Catalogue of the East India Museum in 1857. This discrepancy is explained by the fact of P. Pammon having two females, one closely resembling the male, while the other is totally different from it. A long familiarity with this insect (which replaced by local forms or by closely allied species, occurs in every island of the Archipelago) has convinced me of the correctness of this statement; for in every place where a male allied to P. Pammon is found, a female resembling P. Polytes also occurs, and sometimes, though less frequently than on the continent, another female closely resembling the male: while not only has no male specimen of P. Polytes yet been discovered, but the female (Polytes) has never yet been found in localities to which the male (Pammon) does not extend. In this case, as in the last, distinct species, local forms, and dimorphic specimens, have been confounded under the common appellation of varieties.

But, besides the true P. Polytes, there are several allied forms of females to be considered, namely, P. Theseus (Cramer), P. Melanides (De Haan), P. Elyros (G. R. Gray), and P. Romulus (Linnæus). The dark female figured by Cramer as P. Theseus seems to be the common and perhaps the only form in Sumatra, whereas in Java, Borneo, and Timor, along with males quite identical with those of Sumatra, occur females of the Polytes form, although a single specimen of the true P. Theseus taken at Lombock would seem to show that the two forms do occur together. In the allied species found in the Philippine Islands (P. Alphenor, Cramer = P. Ledebouria, Eschscholtz, the female of which is P. Elyros, G. R. Gray,) forms corresponding to these extremes occur, along with a number of intermediate varieties, as shown by a fine series in the British Museum. We have here an indication of how dimorphism may be produced; for let the extreme Philippine forms be better suited to their conditions of existence than the intermediate connecting links, and the latter will gradually die out, leaving two distinct forms of the same insect, each adapted to some special conditions. As these conditions are sure to vary in different districts, it will often happen, as in Sumatra and Java, that the one form will predominate in the one island, the other in the adjacent one. In the island of Borneo there seems to be a third form; for P. Melanides (De Haan) evidently belongs to this group, and has all the chief characteristics of P. Theseus, with a modified colouration of the hind wings. I now come to an insect which, if I am correct, offers one of the most interesting cases of variation yet adduced. Papilio Romulus, a butterfly found over a large part of India and Ceylon, and not uncommon in collections, has always been considered a true and independent species, and no suspicions have been expressed regarding it. But a male of this form does not, I believe, exist. I have examined the fine series in the British Museum, in the East India Company's Museum, in the Hope Museum at Oxford, in Mr. Hewitson's and several other private collections, and can find nothing but females; and for this common butterfly no male partner can be found except the equally common P. Pammon, a species already provided with two wives, and yet to whom we shall be forced, I believe, to assign a third. On carefully examining P. Romulus, I find that in all essential characters--the form and texture of the wings, the length of the antennæ, the spotting of the head and thorax, and even the peculiar tints and shades with which it is ornamented--it corresponds exactly with the other females of the Pammon group; and though, from the peculiar marking of the fore wings, it has at first sight a very different aspect, yet a closer examination shows that every one of its markings could be produced by slight and almost imperceptible modifications of the various allied forms. I fully believe, therefore, that I shall be correct in placing P. Romulus as a third Indian form of the female P. Pammon, corresponding to P. Melanides, the third form of the Malayan P. Theseus. I may mention here that the females of this group have a superficial resemblance to the Polydorus group of Papilios, as shown by P. Theseus having been considered to be the female of P. Antiphus, and by P. Romulus being arranged next to P. Hector. There is no close affinity between these two groups of Papilio, and I am disposed to believe that we have here a case of mimicry, brought about by the same causes which Mr. Bates has so well explained in his account of the Heliconidæ, and which has led to the singular exuberance of polymorphic forms in this and allied groups of the genus Papilio. I shall have to devote a section of my essay to the consideration of this subject.

The third example of polymorphism I have to bring forward is Papilio Ormenus, which is closely allied to the well-known P. Erechtheus, of Australia. The most common form of the female also resembles that of P. Erechtheus; but a totally different-looking insect was found by myself in the Aru Islands, and figured by Mr. Hewitson under the name of P. Onesimus, which subsequent observation has convinced me is a second form of the female of P. Ormenus. Comparison of this with Boisduval's description of P. Amanga, a specimen of which from New Guinea is in the Paris Museum, shows the latter to be a closely similar form; and two other specimens were obtained by myself, one in the island of Goram and the other in Waigiou, all evidently local modifications of the same form. In each of these localities males and ordinary females of P. Ormenus were also found. So far there is no evidence that these light-coloured insects are not females of a distinct species, the males of which have not been discovered. But two facts have convinced me this is not the case. At Dorey, in New Guinea, where males and ordinary females closely allied to P. Ormenus occur (but which seem to me worthy of being separated as a distinct species), I found one of these light-coloured females closely followed in her flight by three males, exactly in the same manner as occurs (and, I believe, occurs only) with the sexes of the same species. After watching them a considerable time, I captured the whole of them, and became satisfied that I had discovered the true relations of this anomalous form. The next year I had corroborative proof of the correctness of this opinion by the discovery in the island of Batchian of a new species allied to P. Ormenus, all the females of which, either seen or captured by me, were of one form, and much more closely resembling the abnormal light-coloured females of P. Ormenus and P. Pandion than the ordinary specimens of that sex. Every naturalist will, I think, agree that this is strongly confirmative of the supposition that both forms of female are of one species; and when we consider, further, that in four separate islands, in each of which I resided for several months, the two forms of female were obtained and only one form of male ever seen, and that about the same time, M. Montrouzier in Woodlark Island, at the other extremity of New Guinea (where he resided several years, and must have obtained all the large Lepidoptera of the island), obtained females closely resembling mine, which, in despair at finding no appropriate partners for them, he mates with a widely different species--it becomes, I think, sufficiently evident this is another case of polymorphism of the same nature as those already pointed out in P. Pammon and P. Memnon. This species, however, is not only dimorphic, but trimorphic; for, in the island of Waigiou, I obtained a third female quite distinct from either of the others, and in some degree intermediate between the ordinary female and the male. The specimen is particularly interesting to those who believe, with Mr. Darwin, that extreme difference of the sexes has been gradually produced by what he terms sexual selection, since it may be supposed to exhibit one of the intermediate steps in that process, which has been accidentally preserved in company with its more favoured rivals, though its extreme rarity (only one specimen having been seen to many hundreds of the other form) would indicate that it may soon become extinct.

The only other case of polymorphism in the genus Papilio, at all equal in interest to those I have now brought forward, occurs in America; and we have, fortunately, accurate information about it. Papilio Turnus is common over almost the whole of temperate North America; and the female resembles the male very closely. A totally different-looking insect both in form and colour, Papilio Glaucus, inhabits the same region; and though, down to the time when Boisduval published his "Species Général," no connexion was supposed to exist between the two species, it is now well ascertained that P. Glaucus is a second female form of P. Turnus. In the "Proceedings of the Entomological Society of Philadelphia," Jan., 1863, Mr. Walsh gives a very interesting account of the distribution of this species. He tells us that in the New England States and in New York all the females are yellow, while in Illinois and further south all are black; in the intermediate region both black and yellow females occur in varying proportions. Lat. 37° is approximately the southern limit of the yellow form, and 42° the northern limit of the black form; and, to render the proof complete, both black and yellow insects have been bred from a single batch of eggs. He further states that, out of thousands of specimens, he has never seen or heard of intermediate varieties between these forms. In this interesting example we see the effects of latitude in determining the proportions in which the individuals of each form should exist. The conditions are *here* favourable to the one form, *there* to the other; but we are by no means to suppose that these conditions consist in climate alone. It is highly probable that the existence of enemies, and of competing forms of life, may be the main determining influences; and it is much to be wished that such a competent observer as Mr. Walsh would endeavour to ascertain what are the adverse causes which are most efficient in keeping down the numbers of each of these contrasted forms.

Dimorphism of this kind in the animal kingdom does not seem to have any direct relations to the reproductive powers, as Mr. Darwin has shown to be the case in plants, nor does it appear to be very general. One other case only is known to me in another family of my eastern Lepidoptera, the Pieridæ; and but few occur in the Lepidoptera of other countries. The spring and autumn broods of some European species differ very remarkably; and this must be considered as a phenomenon of an analogous though not of an identical nature, while the Araschnia prorsa, of Central Europe, is a striking example of this alternate or seasonal dimorphism. Among our nocturnal Lepidoptera, I am informed, many analogous cases occur; and as the whole history of many of these has been investigated by breeding successive generations from the egg, it is to be hoped that some of our British Lepidopterists will give us a connected account of all the abnormal phenomena which they present. Among the Coleoptera Mr. Pascoe has pointed out the existence of two forms of the male sex in seven species of the two genera Xenocerus and Mecocerus belonging to the family Anthribidæ, (Proc. Ent. Soc. Lond., 1862); and no less than six European Water-beetles, of the genus Dytiscus, have females of two forms, the most common having the elytra deeply sulcate, the rarer smooth as in the males. The three, and sometimes four or more, forms under which many Hymenopterous insects (especially Ants) occur, must be considered as a related phenomenon, though here each form is specialized to a distinct function in the economy of the species. Among the higher animals, albinoism and melanism may, as I have already stated, be considered as analogous facts; and I met with one case of a bird, a species of Lory (Eos fuscata), clearly existing under two differently coloured forms, since I obtained both sexes of each from a single flock, while no intermediate specimens have yet been found.

The fact of the two sexes of one species differing very considerably is so common, that it attracted but little attention till Mr. Darwin showed how it could in many cases be explained by the principle of sexual selection. For instance, in most polygamous animals the males fight for the possession of the females, and the victors, always becoming the progenitors of the succeeding generation, impress upon their male offspring their own superior size, strength, or unusually developed offensive weapons. It is thus that we can account for the spurs and the superior strength and size of the males in Gallinaceous birds, and also for the large canine tusks in the males of fruit-eating Apes. So the superior beauty of plumage and special adornments of the males of so many birds can be explained by supposing (what there are many facts to prove) that the females prefer the most beautiful and perfect-plumaged males, and that thus, slight accidental variations of form and colour have been accumulated, till they have produced the wonderful train of the Peacock and the gorgeous plumage of the Bird of Paradise. Both these causes have no doubt acted partially in insects, so many species possessing horns and powerful jaws in the male sex only, and still more frequently the males alone rejoicing in rich colours or sparkling lustre. But there is here another cause which has led to sexual differences, viz., a special adaptation of the sexes to diverse habits or modes of life. This is well seen in female Butterflies (which are generally weaker and of slower flight), often having colours better adapted to concealment; and in certain South American species (Papilio torquatus) the females, which inhabit the forests, resemble the Æneas group of Papilios which abound in similar localities, while the males, which frequent the sunny open river-banks, have a totally different colouration. In these cases, therefore, natural selection seems to have acted independently of sexual selection; and all such cases may be considered as examples of the simplest dimorphism, since the offspring never offer intermediate varieties between the parent forms.

The phenomena of dimorphism and polymorphism may be well illustrated by supposing that a blue-eyed, flaxen-haired Saxon man had two wives, one a black-haired, red-skinned Indian squaw, the other a woolly-headed, sooty-skinned negress--and that instead of the children being mulattoes of brown or dusky tints, mingling the separate characteristics of their parents in varying degrees, all the boys should be pure Saxon boys like their father, while the girls should altogether resemble their mothers. This would be thought a sufficiently wonderful fact; yet the phenomena here brought forward as existing in the insect-world are still more extraordinary; for each mother is capable not only of producing male offspring like the father, and female like herself, but also of producing other females exactly like her fellow-wife, and altogether differing from herself. If an island could be stocked with a colony of human beings having similar physiological idiosyncrasies with Papilio Pammon or Papilio Ormenus, we should see white men living with yellow, red, and black women, and their offspring always reproducing the same types; so that at the end of many generations the men would remain pure white, and the women of the same well-marked races as at the commencement.

The distinctive character therefore of dimorphism is this, that the union of these distinct forms does not produce intermediate varieties, but reproduces the distinct forms unchanged. In simple varieties, on the other hand, as well as when distinct local forms or distinct species are crossed, the offspring never resembles either parent exactly, but is more or less intermediate between them. Dimorphism is thus seen to be a specialized result of variation, by which new physiological phenomena have been developed; the two should therefore, whenever possible, be kept separate.

3. *Local form, or variety.*--This is the first step in the transition from variety to species. It occurs in species of wide range, when groups of individuals have become partially isolated in several points of its area of distribution, in each of which a characteristic form has become more or less completely segregated. Such forms are very common in all parts of the world, and have often been classed by one author as varieties, by another as species. I restrict the term to those cases where the difference of the forms is very slight, or where the segregation is more or less imperfect. The best example in the present group is Papilio Agamemnon, a species which ranges over the greater part of tropical Asia, the whole of the Malay archipelago, and a portion of the Australian and

Pacific regions. The modifications are principally of size and form, and, though slight, are tolerably constant in each locality. The steps, however, are so numerous and gradual that it would be impossible to define many of them, though the extreme forms are sufficiently distinct. Papilio Sarpedon presents somewhat similar but less numerous variations.

4. *Co-existing Variety.*--This is a somewhat doubtful case. It is when a slight but permanent and hereditary modification of form exists in company with the parent or typical form, without presenting those intermediate gradations which would constitute it a case of simple variability. It is evidently only by direct evidence of the two forms breeding separately that this can be distinguished from dimorphism. The difficulty occurs in Papilio Jason, and P. Evemon, which inhabit the same localities, and are almost exactly alike in form, size, and colouration, except that the latter always wants a very conspicuous red spot on the under surface, which is found not only in P. Jason, but in all the allied species. It is only by breeding the two insects that it can be determined whether this is a case of a co-existing variety or of dimorphism. In the former case, however, the difference being constant and so very conspicuous and easily defined, I see not how we could escape considering it as a distinct species. A true case of co-existing forms would, I consider, be produced, if a slight variety had become fixed as a local form, and afterwards been brought into contact with the

parent species, with little or no intermixture of the two; and such instances do very probably occur.

5. *Race or subspecies.*--These are local forms completely fixed and isolated; and there is no possible test but individual opinion to determine which of them shall be considered as species and which varieties. If stability of form and "*the constant transmission of some characteristic peculiarity of organization*" is the test of a species (and I can find no other test that is more certain than individual opinion) then every one of these fixed races, confined as they almost always are to distinct and limited areas, must be regarded as a species; and as such I have in most cases treated them. The various modifications of Papilio Ulysses, P. Peranthus, P. Codrus, P. Eurypilus, P. Helenus, &c., are excellent examples; for while some present great and well-marked, others offer slight and inconspicuous differences, yet in all cases these differences seem equally fixed and permanent. If, therefore, we call some of these forms species, and others varieties, we introduce a purely arbitrary distinction, and shall never be able to decide where to draw the line. The races of Papilio Ulysses, for example, vary in amount of modification from the scarcely differing New Guinea form to those of Woodlark Island and New Caledonia, but all seem equally constant; and as most of these had already been named and described as species, I have added the New Guinea form under the name of P. Autolycus. We thus get a little group of Ulyssine Papilios,

the whole comprised within a very limited area, each one confined to a separate portion of that area, and, though differing in various amounts, each apparently constant. Few naturalists will doubt that all these may and probably have been derived from a common stock, and therefore it seems desirable that there should be a unity in our method of treating them; either call them all *varieties* or all *species.* Varieties, however, continually get overlooked; in lists of species they are often altogether unrecorded; and thus we are in danger of neglecting the interesting phenomena of variation and distribution which they present. I think it advisable, therefore, to name all such forms; and those who will not accept them as species may consider them as subspecies or races.

6. *Species.*--Species are merely those strongly marked races or local forms which when in contact do not intermix, and when inhabiting distinct areas are generally believed to have had a separate origin, and to be incapable of producing a fertile hybrid offspring. But as the test of hybridity cannot be applied in one case in ten thousand, and even if it could be applied would prove nothing, since it is founded on an assumption of the very question to be decided--and as the test of separate origin is in every case inapplicable--and as, further, the test of non-intermixture is useless, except in those rare cases where the most closely allied species are found inhabiting the same area, it will be evident that we have no

means whatever of distinguishing so-called "true species" from the several modes of variation here pointed out, and into which they so often pass by an insensible gradation. It is quite true that, in the great majority of cases, what we term "species" are so well marked and definite that there is no difference of opinion about them; but as the test of a true theory is, that it accounts for, or at the very least is not inconsistent with, the whole of the phenomena and apparent anomalies of the problem to be solved, it is reasonable to ask that those who deny the origin of species by variation and selection should grapple with the facts in detail, and show how the doctrine of the distinct origin and permanence of species will explain and harmonize them. It has been recently asserted by Dr. J. E. Gray (in the Proceedings of the Zoological Society for 1863, page 134), that the difficulty of limiting species is in proportion to our ignorance, and that just as groups or countries are more accurately known and studied in greater detail the limits of species become settled. This statement has, like many other general assertions, its portion of both truth and error. There is no doubt that many uncertain species, founded on few or isolated specimens, have had their true nature determined by the study of a good series of examples: they have been thereby established as species or as varieties; and the number of times this has occurred is doubtless very great. But there are other, and equally trustworthy cases, in which, not single species, but whole groups have, by the study of a vast accumulation of materials, been proved to have no definite specific limits. A few of these must be adduced. In Dr. Carpenter's "Introduction to the Study of the Foraminifera," he states that *"there is not a single specimen of plant or animal of which the range of variation has been studied by the collocation and comparison of so large a number of specimens as have passed under the review of Messrs. Williamson, Parker, Rupert Jones, and myself, in our studies of the types of this group;"* and the result of this extended comparison of specimens is stated to be, *"The range of variation is so great among the Foraminifera as to include not merely those differential characters which have been usually accounted SPECIFIC, but also those upon which the greater part of the GENERA of this group have been founded, and even in some instances those of its ORDERS"* (Foraminifera, Preface, x). Yet this same group had been divided by D'Orbigny and other authors into a number of clearly defined *families, genera,* and *species,* which these careful and conscientious researches have shown to have been almost all founded on incomplete knowledge.

Professor DeCandolle has recently given the results of an extensive review of the species of Cupuliferæ. He finds that the best-known species of oaks are those which produce most varieties and subvarieties; that they are often surrounded by provisional species; and, with the fullest materials at his command, two-thirds of the species he considers more or less doubtful. His general conclusion is, that "*in botany the lowest series of groups, SUBVARIETIES, VARIETIES, and RACES are very badly limited; these can be grouped into SPECIES a little less vaguely limited, which again can be formed into sufficiently precise GENERA.*" This general conclusion is entirely objected to by the writer of the article in the "Natural History Review," who, however, does not deny its applicability to the particular order under discussion, while this very difference of opinion is another proof that difficulties in the determination of species do not, any more than in the higher groups, vanish with increasing materials and more accurate research.

Another striking example of the same kind is seen in the genera Rubus and Rosa, adduced by Mr. Darwin himself; for though the amplest materials exist for a knowledge of these groups, and the most careful research has been bestowed upon them, yet the various species have not thereby been accurately limited and defined so as to satisfy the majority of botanists. In Mr. Baker's revision of the British Roses, just published by the Linnæan Society, the author includes under the single species Rosa canina, no less than twenty-eight named *varieties*, distinguished by more or less constant characters and often confined to special localities; and to these are referred about seventy of the *species* of Continental and British botanists.

Dr. Hooker seems to have found the same thing in his

study of the Arctic flora. For though he has had much of the accumulated materials of his predecessors to work upon, he continually expresses himself as unable to do more than group the numerous and apparently fluctuating forms into more or less imperfectly defined species. In his paper on the "Distribution of Arctic Plants," (Trans. Linn. Soc. xxiii.,) Dr. Hooker says:--"The most able and experienced descriptive botanists vary in their estimate of the value of the 'specific term' to a much greater extent than is generally supposed." . . . "I think I may safely affirm that the 'specific term' has three different standard values, all current in descriptive botany, but each more or less confined to one class of observers." . . . "This is no question of what is right or wrong as to the real value of the specific term; I believe each is right according to the standard he assumes as the specific."

Lastly, I will adduce Mr. Bates's researches on the Amazons. During eleven years he accumulated vast materials, and carefully studied the variation and distribution of insects. Yet he has shown that many species of Lepidoptera, which before offered no special difficulties, are in reality most intricately combined in a tangled web of affinities, leading by such gradual steps from the slightest and least stable variations to fixed races and well-marked species, that it is very often impossible to draw those sharp dividing-lines which it is supposed that a careful study and full materials will always enable us to do.

These few examples show, I think, that in every department of nature there occur instances of the instability of specific form, which the increase of materials aggravates rather than diminishes. And it must be remembered that the naturalist is rarely likely to err on the side of imputing greater indefiniteness to species than really exists. There is a completeness and satisfaction to the mind in defining and limiting and naming a species, which leads us all to do so whenever we conscientiously can, and which we know has led many collectors to reject vague intermediate forms as destroying the symmetry of their cabinets. We must therefore consider these cases of excessive variation and instability as being thoroughly well established; and to the objection that, after all, these cases are but few compared with those in which species can be limited and defined, and are therefore merely exceptions to a general rule, I reply that a true law embraces all apparent exceptions, and that to the great laws of nature there are no real exceptions--that what appear to be such are equally results of law, and are often (perhaps indeed always) those very results which are most important as revealing the true nature and action of the law. It is for such reasons that naturalists now look upon the study of *varieties* as more important than that of well-fixed species. It is in the former that we see nature still at work, in the very act of producing those wonderful modifications of form, that endless variety of colour, and that complicated harmony of relations, which gratify every sense and give occupation to every faculty of the true lover of nature.

Variation as specially influenced by Locality.

The phenomena of variation as influenced by locality have not hitherto received much attention. Botanists, it is true, are acquainted with the influences of climate, altitude, and other physical conditions, in modifying the forms and external characteristics of plants; but I am not aware that any peculiar influence has been traced to locality, independent of climate. Almost the only case I can find recorded is mentioned in that repertory of natural-history facts, "The Origin of Species," viz. that herbaceous groups have a tendency to become arboreal in islands. In the animal world, I cannot find that any facts have been pointed out as showing the special influence of locality in giving a peculiar *facies* to the several disconnected species that inhabit it. What I have to adduce on this matter will therefore, I hope, possess some interest and novelty.

On examining the closely allied species, local forms, and varieties distributed over the Indian and Malayan regions, I find that larger or smaller districts, or even single islands, give a special character to the majority of their Papilionidæ. For instance: 1. The species of the Indian region (Sumatra,

Java, and Borneo) are almost invariably smaller than the allied species inhabiting Celebes and the Moluccas; 2. The species of New Guinea and Australia are also, though in a less degree, smaller than the nearest species or varieties of the Moluccas; 3. In the Moluccas themselves the species of Amboyna are the largest; 4. The species of Celebes equal or even surpass in size those of Amboyna; 5. The species and varieties of Celebes possess a striking character in the form of the anterior wings, different from that of the allied species and varieties of all the surrounding islands; 6. Tailed species in India or the Indian region become tailless as they spread eastward through the archipelago; 7. In Amboyna and Ceram the females of several species are dull-coloured, while in the adjacent islands they are more brilliant.

Local variation of Size.--Having preserved the finest and largest specimens of Butterflies in my own collection, and having always taken for comparison the largest specimens of the same sex, I believe that the tables I now give are sufficiently exact. The differences of expanse of wings are in most cases very great, and are much more conspicuous in the specimens themselves than on paper. It will be seen that no less than fourteen Papilionidæ inhabiting Celebes and the Moluccas are from one-third to one-half greater in extent of wing than the allied species representing them in Java, Sumatra, and Borneo. Six species inhabiting Amboyna are larger than the closely allied forms of the northern Moluccas and New Guinea by about one-sixth. These include almost every case in which closely allied species can be compared.

Species of Papilionidæ of the Moluccas and Celebes (large).		Closely allied species of Java and the Indian region (small).	
	Expanse. Inches.		Expanse. Inches.
Ornithoptera Helena Amboyna) ... ... 7·6		O. Pompeus ... ... 5·8	
		O. Amphrisius ... ... 6·0	
Papilio Adamantius (Celebes) ... ... 5·8			
P. Lorquinianus (Moluccas) ... ... 4·8		P. Peranthus ... ... 3·8	

Species of Papilionidæ of the Moluccas and Celebes (large).	Expanse. Inches.	Closely allied species of Java and the Indian region (small).	Expanse. Inches.
P. Blumei (Celebes)	5·4	P. Brama	4·0
P. Alphenor (Celebes)	4·8	P. Theseus	3·6
P. Gigon (Celebes)	5·4	P. Demolion	4·0
P. Deucalion (Celebes)	4·6	P. Macareus	3·7
P. Agamemnon, var. (Celebes)	4·4	P. Agamemnon, var.	3 8
P. Eurypilus (Moluccas)	4·0	P. Jason	3·4
P. Telephus (Celebes)	4·3		
P. Ægisthus (Moluccas)	4·4	P. Rama	3·2
P. Milon (Celebes)	4·4	P. Sarpedon	3·8
P. Androcles (Celebes)	4·8	P. Antiphates	3·7
P. Polyphontes (Celebes)	4·6	P. Diphilus	3·9
Leptocircus Ennius (Celebes)	2·0	L. Meges	1·8

Species inhabiting Amboyna (large).		Allied species of New Guinea and the North Moluccas (smaller).	
Papilio Ulysses	6·1	P. Autolycus	5 2
		P. Telegonus	4·0
P. Polydorus	4·9	P. Leodamas	4·0
P. Deiphobus	6·8	P. Deiphontes	5·8
P. Gambrisius	6·4	P. Ormenus	5·6
		P. Tydeus	6·0
P. Codrus	5·1	P. Codrus, var. papuensis	4·3
Ornithoptera Priamus, (male)	8·3	Ornithoptera Poseidon, (male)	7·0

Local variation of Form.--The differences of form are equally clear. Papilio Pammon everywhere on the continent is tailed in both sexes. In Java, Sumatra, and Borneo, the closely allied P. Theseus has a very short tail, or tooth only,

48

in the male, while in the females the tail is retained. Further east, in Celebes and the South Moluccas, the hardly separable P. Alphenor has quite lost the tail in the male, while the female retains it, but in a narrower and less spatulate form. A little further, in Gilolo, P. Nicanor has completely lost the tail in both sexes.

Papilio Agamemnon exhibits a somewhat similar series of changes. In India it is always tailed; in the greater part of the archipelago it has a very short tail; while far east, in New Guinea and the adjacent islands, the tail has almost entirely disappeared.

In the Polydorus-group two species, P. Antiphus and P. Diphilus, inhabiting India and the Indian region, are tailed, while the two which take their place in the Moluccas, New Guinea, and Australia, P. Polydorus and P. Leodamas, are destitute of tail, the species furthest east having lost this ornament the most completely.

Western species, Tailed.	*Allied Eastern species not Tailed.*
Papilio Pammon (India)	P. Thesus (Islands) minute tail.
P. Agamemnon, var. (India)	P. Agamemnon, var. (Islands).
P. Antiphus (India, Java)	P. Polydorus (Moluccas).
P. Diphilus (India, Java)	P. Leodamas (New Guinea).

The most conspicuous instance of local modification of form, however, is exhibited in the island of Celebes, which in this respect, as in some others, stands alone and isolated in the whole archipelago. Almost every species of Papilio inhabiting Celebes has the wings of a peculiar shape, which distinguishes them at a glance from the allied species of every other island. This peculiarity consists, first, in the upper wings being generally more elongate and falcate; and secondly, in the costa or anterior margin being much more curved, and in most instances exhibiting near the base an abrupt bend or elbow, which in some species is very conspicuous. This peculiarity is visible, not only when the Celebesian species are compared with their small-sized allies of Java and Borneo, but also, and in an almost equal degree, when the large forms of Amboyna and the Moluccas are the objects of comparison, showing that this is quite a distinct phenomenon from the difference of size which has just been pointed out.

In the following Table I have arranged the chief Papilios of Celebes in the order in which they exhibit this characteristic form most prominently.

Papilios of Celebes, having the wings falcate or with abruptly curved costa.	*Closely allied Papilios of the surrounding islands, with less falcate wings and slightly curved costa.*
1. P. Gigon	P. Demolion (Java).
2. P. Pamphylus	P. Jason (Sumatra).
3. P. Milon	P. Sarpedon (Moluccas, Java).
4. P. Agamemnon, var.	P. Agamemnon, var. (Borneo).
5. P. Adamantius	P. Peranthus (Java).
6. P. Ascalaphus	P. Deiphontes (Gilolo).

7. P. Sataspes	P. Helenus (Java).
8. P. Blumei	P. Brama (Sumatra).
9. P. Androcles	P. Antiphates (Borneo).
10. P. Rhesus	P. Aristæus (Moluccas).
11. P. Theseus, var. (male)	P. Thesus (male) (Java).
12. P. Codrus, var.	P. Codrus (Moluccas).
13. P. Encelades	P. Leucothoë (Malacca).

It thus appears that every species of Papilio exhibits this peculiar form in a greater or less degree, except one, P. Polyphontes, allied to P. Diphilus of India and P. Polydorus of the Moluccas. This fact I shall recur to again, as I think it helps us to understand something of the causes that may have brought about the phenomenon we are considering. Neither do the genera Ornithoptera and Leptocircus exhibit any traces of this peculiar form. In several other families of Butterflies this characteristic form reappears in a few species. In the Pieridæ the following species, all peculiar to Celebes, exhibit it distinctly:--

1. Pieris Eperia, compared with P. Coronis (Java).

2. Thyca Zebuda, compared with Thyca Descombesi (India).

3. T. Rosenbergii, compared with T. Hyparete (Java).

4. Tachyris Hombronii, compared with T. Lyncida.

5. T. Lycaste, compared with T. Lyncida.

6. T. Zarinda, compared with T. Nero (Malacca).

7. T. Ithome, compared with T. Nephele.

8. Eronia tritæa, compared with Eronia Valeria (Java).

9. Iphias Glaucippe, var., compared with Iphias Glaucippe (Java).

The species of Terias, one or two Pieris, and the genus Callidryas do not exhibit any perceptible change of form.

In the other families there are but few similar examples. The following are all that I can find in my collection:--

Cethosia Æole, compared with Cethosia Biblis (Java). Eurhinia megalonice, compared with Eurhinia Polynice (Borneo).

Limenitis Limire, compared with Limenitis Procris (Java). Cynthia Arsinoë, var., compared with Cynthia Arsinoë (Java, Sumatra, Borneo)

All these belong to the family of the Nymphalidæ. Many other genera of this family, as Diadema, Adolias, Charaxes, and Cyrestis, as well as the entire families of the Danaidæ, Satyridæ, Lycænidæ, and Hesperidæ, present no examples of this peculiar form of the upper wing in the Celebesian species.

Local variations of Colour.--In Amboyna and Ceram the female of the large and handsome Ornithoptera Helena has the large patch on the hind wings constantly of a pale dull ochre or buff colour, while in the scarcely distinguishable varieties from the adjacent islands of Bouru and New Guinea, it is of a golden yellow, hardly inferior in brilliancy to its colour in the male sex. The female of Ornithoptera Priamus (inhabiting Amboyna and Ceram exclusively) is of a pale dusky brown tint, while in all the allied species the same sex is nearly black with contrasted white markings. As a third example, the female of Papilio Ulysses has the blue colour obscured by dull and dusky tints, while in the closely allied species from the surrounding islands, the females are of almost as brilliant an azure blue as the males. A parallel case to this is the occurrence, in the small islands of Goram, Matabello, Ké, and Aru, of several distinct species of Euplæa and Diadema, having broad bands or patches of white, which do not exist in any of the allied species from the larger islands. These facts seem to indicate some local influence in modifying colour, as unintelligible and almost as remarkable as that which has resulted in the modifications of form previously described.

Remarks on the facts of Local variation.

The facts now brought forward seem to me of the highest interest. We see that almost all the species in two important families of the Lepidoptera (Papilionidæ and Pieridæ) acquire, in a single island, a characteristic modification of form distinguishing them from the allied species and varieties of all the surrounding islands. In other equally extensive families no such change occurs, except in one or two isolated species. However we may account for these phenomena, or whether we may be quite unable to account for them, they furnish, in my opinion, a strong corroborative testimony in favour of the doctrine of the origin of species by successive small variations; for we have here slight varieties, local races, and undoubted species, all modified in exactly the same manner, indicating plainly a common cause producing identical results. On the generally received theory of the original distinctness and permanence of species, we are met by this difficulty: one portion of these curiously modified forms are admitted to have been produced by variation and some natural action of local conditions; whilst the other portion, differing from the former only in degree, and connected with them by insensible gradations, are said to have possessed this peculiarity of form at their first creation, or to have derived it from unknown causes of a totally distinct nature. Is not the *à priori* evidence in favour of an identity of the causes that have produced such similar results? and have we not a right to call upon our opponents for some proofs of their own doctrine, and for an explanation of its difficulties, instead of their assuming that they are right, and laying upon us the burthen of disproof?

Let us now see if the facts in question do not themselves furnish some clue to their explanation. Mr. Bates has shown that certain groups of butterflies have a defence against insectivorous animals, independent of swiftness of motion. These are generally very abundant, slow, and weak fliers, and are more or less the objects of mimicry by other groups, which thus gain an advantage in a freedom from persecution similar to that enjoyed by those they resemble. Now the only Papilios which have not in Celebes acquired the peculiar form of wing, belong to a group which is imitated both by other species of Papilio and by Moths of the genus Epicopeia. This group is of weak and slow flight; and we may therefore fairly conclude that it possesses some means of defence (probably in a peculiar odour or taste) which saves it from attack. Now the arched costa and falcate form of wing is generally supposed to give increased powers of flight, or, as seems to me more probable, greater facility in making sudden turnings, and thus baffling a pursuer. But the members of the Polydorus-group (to which belongs the only unchanged Celebesian Papilio), being already guarded against attack, have no need of this increased power of wing; and "natural selection" would therefore have no tendency to produce it. The whole family of Danaidæ are in the same position: they are slow and weak fliers; yet they abound in species and individuals, and are the objects of mimicry. The Satyridæ have also probably a means of protection--perhaps their keeping always near the ground and their generally obscure colours; while the Lycænidæ and Hesperidæ may find security in their small size and rapid motions. In the extensive family of the Nymphalidæ, however, we find that several of the larger species, of comparatively feeble structure, have their wings modified (Cethosia, Limenitis, Junonia, Cynthia), while the large-bodied powerful species, which have all an excessively rapid flight, have exactly the same form of wing in Celebes as in the other islands. On the whole, therefore, we may say that all the butterflies of rather large size, conspicuous colours, and not very swift flight have been affected in the manner described, while the smaller sized and obscure groups, as well as those which are the objects of mimicry, and also those of exceedingly swift flight have remained unaffected.

It would thus appear as if there must be (or once have been) in the island of Celebes, some peculiar enemy to these larger-sized butterflies which does not exist, or is less abundant, in the surrounding islands. Increased powers of flight, or rapidity of turning, was advantageous in baffling this enemy; and the peculiar form of wing necessary to give this would be readily acquired by the action of "natural selection" on the slight variations of form that are continually occurring.

Such an enemy one would naturally suppose to be an insectivorous bird; but it is a remarkable fact that most of the genera of Fly-catchers of Borneo and Java on the one side (Muscipeta, Philentoma,) and of the Moluccas on the other (Monarcha, Rhipidura), are almost entirely absent from Celebes. Their place seems to be supplied by the Caterpillar-catchers (Graucalus, Campephaga, &c.), of which six or seven species are known from Celebes and are very numerous in individuals. We have no positive evidence that these birds pursue butterflies on the wing, but it is highly probable that they do so when other food is scarce. Mr. Bates has suggested to me that the larger Dragon-flies (Æshna, &c.) prey upon butterflies; but I did not notice that they were more abundant in Celebes than elsewhere. However this may be, the fauna of Celebes is undoubtedly highly peculiar in every department of which we have any accurate knowledge; and though we may not be able satisfactorily to trace how it has been effected, there can, I think, be little doubt that the singular modification in the wings of so many of the butterflies of that island is an effect of that complicated action and reaction of all living things upon each other in the struggle for existence, which continually tends to readjust disturbed relations, and to bring every species into harmony with the varying conditions of the surrounding universe.

But even the conjectural explanation now given fails us in the other cases of local modification. Why the species of the Western islands should be smaller than those further east,--why those of Amboyna should exceed in size those of Gilolo and New Guinea--why the tailed species of India should begin to lose that appendage in the islands, and retain no trace of it on the borders of the Pacific,--and why, in three separate cases, the females of Amboyna species should be less gaily attired than the corresponding females of the surrounding islands,--are questions which we cannot at present attempt to answer. That they depend, however, on some general principle is certain, because analogous facts have been observed in other parts of the world. Mr. Bates informs me that, in three distinct groups, Papilios which on the Upper Amazon and in most other parts of South America have spotless upper wings obtain pale or white spots at Pará and on the Lower Amazon; and also that the Æneas-group of Papilios never have tails in the equatorial regions and the Amazons valley, but gradually acquire tails in many cases as they range towards the northern or southern tropic. Even in Europe we have somewhat similar facts; for the species and varieties of butterflies peculiar to the island of Sardinia are generally smaller and more deeply coloured than those of the mainland, and the same has recently been shown to be the case with the common tortoiseshell butterfly in the Isle of Man; while Papilio Hospiton, peculiar to the former island, has lost the tail, which is a prominent feature of the closely allied P. Machaon.

Facts of a similar nature to those now brought forward would no doubt be found to occur in other groups of insects, were local faunas carefully studied in relation to those of the surrounding countries; and they seem to indicate that climate and other physical causes have, in some cases, a very powerful effect in modifying specific form and colour, and thus directly aid in producing the endless variety of nature.

Mimicry.

Having fully discussed this subject in the preceding essay, I have only to adduce such illustrations of it, as are furnished by the Eastern Papilionidæ, and to show their bearing upon the phenomena of variation already mentioned. As in America, so in the Old World, species of Danaidæ are the objects which the other families most often imitate. But besides these, some genera of Morphidæ and one section of the genus Papilio are also less frequently copied. Many species of Papilio mimic other species of these three groups so closely that they are undistinguishable when on the wing; and in every case the pairs which resemble each other inhabit the same locality.

The following list exhibits the most important and best marked cases of mimicry which occur among the Papilionidæ of the Malayan region and India:--

Mimickers.	Species mimicked.	Common habitat.

DANAIDÆ.

1. Papilio paradoxa (male & female)	Euplœa Midamus (male & female)	Sumatra, &c.
2. P. Caunus	E. Rhadamanthus	Borneo & Sumatra.
3. P. Thule	Danais sobrina	New Guinea.
4. P. Macareus	D. Aglaia	Malacca, Java.
5. Papilio Agestor	Danais Tytia	Northern India.
6. P. Idæoides	Hestia Leuconoë	Philippines.
7. P. Delessertii	Ideopsis daos	Penang.

MORPHIDÆ.

8. P. Pandion (female)	Drusilla bioculata	New Guinea.

PAPILIO (POLYDORUS- and COON-groups).

9. P. Pammon (Romulus, female)	Papilio Hector	India.
10. P. Theseus, var. (female)	P. Antiphus	Sumatra, Borneo.
11. P. Theseus, var. (female)	P. Diphilus	Sumatra, Java.
12. P. Memnon, var. (Achates, female)	P. Coon	Sumatra.
13. P. Androgeus, var. (Achates, female)	P. Doubledayi	Northern India.
14. P. Œnomaus (female)	P. Liris	Timor.

We have, therefore, fourteen species or marked varieties of Papilio, which so closely resemble species of other groups in their respective localities, that it is not possible to impute the resemblance to accident. The first two in the list (Papilio paradoxa and P. Caunus) are so exactly like Euplœa Midamus and E. Rhadamanthus on the wing, that although they fly very slowly, I was quite unable to distinguish them. The first is a very interesting case, because the male and female differ considerably, and each mimics the corresponding sex of the Euplœa. A new species of Papilio which I discovered in New Guinea resembles Danais sobrina, from the same country, just as Papilio Marcareus resembles Danais Aglaia in Malacca, and (according to Dr. Horsfield's figure) still more closely in Java. The Indian Papilio Agestor closely imitates Danais Tytia, which has quite a different style of colouring from the preceding; and the extraordinary Papilio Idæoides from the Philippine Islands, must, when on the wing, perfectly resemble the Hestia Leuconoë of the same region, as also does the Papilio Delessertii imitate the Ideopsis daos from Penang. Now in every one of these cases the Papilios are very scarce, while the Danaidæ which they resemble are exceedingly abundant--most of them swarming so as to be a positive nuisance to the collecting entomologist by continually hovering before him when he is in search of newer and more varied captures. Every garden, every roadside, the suburbs of every village are full of them, indicating very clearly that their life is an easy one, and that they are free from persecution by the foes which keep down the population of less favoured races. This superabundant population has been shown by Mr. Bates to be a general characteristic of all American groups and species which are objects of mimicry; and it is interesting to find his observations confirmed by examples on the other side of the globe.

The remarkable genus Drusilla, a group of pale-coloured butterflies, more or less adorned with ocellate spots, is also the object of mimicry by three distinct genera (Melanitis, Hyantis, and Papilio). These insects, like the Danaidæ, are abundant in individuals, have a very weak and slow flight, and do not seek concealment, or appear to have any means of protection from insectivorous creatures. It is natural to conclude, therefore, that they have some hidden property which saves them from attack; and it is easy to see that when any other insects, by what we call accidental variation, come more or less remotely to resemble them, the latter will share to some extent in their immunity. An extraordinary dimorphic form of the female of Papilio Ormenus has come to resemble the Drusillas sufficiently to be taken for one of that group at a little distance; and it is curious that I captured one of these Papilios in the Aru Islands hovering along the ground, and settling on it occasionally, just as it is the habit of the Drusillas to do. The resemblance in this case is only general; but this form of Papilio varies much, and there is therefore material for natural selection to act upon, so as ultimately to produce a copy as exact as in the other cases.

The eastern Papilios allied to Polydorus, Coon, and Philoxenus, form a natural section of the genus resembling, in many respects, the Æneas-group of South America, which they may be said to represent in the East. Like them, they are forest insects, have a low and weak flight, and in their favourite localities are rather abundant in individuals; and like them, too, they are the objects of mimicry. We may conclude, therefore, that they possess some hidden means of protection, which makes it useful to other insects to be mistaken for them.

The Papilios which resemble them belong to a very distinct section of the genus, in which the sexes differ greatly; and it is those females only which differ most from the males, and which have already been alluded to as exhibiting instances of dimorphism, which resemble species of the other group.

The resemblance of P. Romulus to P. Hector is, in some specimens, very considerable, and has led to the two species being placed following each other in the British Museum Catalogues and by Mr. E. Doubleday. I have shown, however, that P. Romulus is probably a dimorphic form of the female P. Pammon, and belongs to a distinct section of the genus.

The next pair, Papilio Theseus, and P. Antiphus, have been united as one species both by De Haan and in the British Museum Catalogues. The ordinary variety of P. Theseus found in Java almost as nearly resembles P.

Diphilus, inhabiting the same country. The most interesting case, however, is the extreme female form of P. Memnon (figured by Cramer under the name of P. Achates), which has acquired the general form and markings of P. Coon, an insect which differs from the ordinary male P. Memnon, as much as any two species which can be chosen in this extensive and highly varied genus; and, as if to show that this resemblance is not accidental, but is the result of law, when in India we find a species closely allied to P. Coon, but with red instead of yellow spots (P. Doubledayi), the corresponding variety of P. Androgeus (P. Achates, Cramer, 182, A, B,) has acquired exactly the same peculiarity of having red spots instead of yellow. Lastly, in the island of Timor, the female of P. Œnomaus (a species allied to P. Memnon) resembles so closely P. Liris (one of the Polydorus-group), that the two, which were often seen flying together, could only be distinguished by a minute comparison after being captured.

The last six cases of mimicry are especially instructive, because they seem to indicate one of the processes by which dimorphic forms have been produced. When, as in these cases, one sex differs much from the other, and varies greatly itself, it may happen that occasionally individual variations will occur having a distant resemblance to groups which are the objects of mimicry, and which it is therefore advantageous to resemble. Such a variety will have a better chance of preservation; the individuals possessing it will be multiplied;

and their accidental likeness to the favoured group will be rendered permanent by hereditary transmission, and, each successive variation which increases the resemblance being preserved, and all variations departing from the favoured type having less chance of preservation, there will in time result those singular cases of two or more isolated and fixed forms, bound together by that intimate relationship which constitutes them the sexes of a single species. The reason why the females are more subject to this kind of modification than the males is, probably, that their slower flight, when laden with eggs, and their exposure to attack while in the act of depositing their eggs upon leaves, render it especially advantageous for them to have some additional protection. This they at once obtain by acquiring a resemblance to other species which, from whatever cause, enjoy a comparative immunity from persecution.

Concluding remarks on Variation in Lepidoptera.

This summary of the more interesting phenomena of variation presented by the eastern Papilionidæ is, I think, sufficient to substantiate my position, that the Lepidoptera are a group that offer especial facilities for such inquiries; and it will also show that they have undergone an amount of special adaptive modification rarely equalled among the more highly organized animals. And, among the Lepidoptera, the great and pre-eminently tropical families of Papilionidæ and Danaidæ seem to be those in which complicated adaptations to the surrounding organic and inorganic universe have been most completely developed, offering in this respect a striking analogy to the equally extraordinary, though totally different, adaptations which present themselves in the Orchideæ, the only family of plants in which mimicry of other organisms appears to play any important part, and the only one in which cases of conspicuous polymorphism occur; for as such we must class the male, female, and hermaphrodite forms of Catasetum tridentatum, which differ so greatly in form and structure that they were long considered to belong to three distinct genera.

Arrangement and Geographical Distribution of the Malayan Papilionidæ.

Arrangement.--Although the species of Papilionidæ inhabiting the Malayan region are very numerous, they all belong to three out of the nine genera into which the family is divided. One of the remaining genera (Eurycus) is restricted to Australia, and another (Teinopalpus) to the Himalayan Mountains, while no less than four (Parnassius, Doritis, Thais, and Sericinus) are confined to Southern

Europe and to the mountain-ranges of the Palæarctic region. The genera Ornithoptera and Leptocircus are highly characteristic of Malayan entomology, but are uniform in character and of small extent. The genus Papilio, on the other hand, presents a great variety of forms, and is so richly represented in the Malay Islands, that more than one-fourth of all the known species are found there. It becomes necessary, therefore, to divide this genus into natural groups before we can successfully study its geographical distribution.

Owing principally to Dr. Horsfield's observations in Java, we are acquainted with a considerable number of the larvæ of Papilios; and these furnish good characters for the primary division of the genus into natural groups. The manner in which the hinder wings are plaited or folded back at the abdominal margin, the size of the anal valves, the structure of the antennæ, and the form of the wings are also of much service, as well as the character of the flight and the style of colouration. Using these characters, I divide the Malayan Papilios into four sections, and seventeen groups, as follows:--

Genus ORNITHOPTERA.

a. Priamus-group. ⎫
c. Brookeanus-group. ⎬ Black and green.

b. Pompeus-group. Black and yellow.

Genus PAPILIO.

A. Larvæ short, thick, with numerous fleshy tubercles; of a purplish colour.

  a. Nox-group. Abdominal fold in male very large; anal valves small, but swollen; antennæ moderate; wings entire, or tailed; includes the Indian Philoxenus-group.

  b. Coon-group. Abdominal fold in male small; anal valves small, but swollen; antennæ moderate; wings tailed.

  c. Polydorus-group. Abdominal fold in male small, or none; anal valves small or obsolete, hairy; wings tailed or entire.

B. Larvæ with third segment swollen, transversely or obliquely banded; pupa much bent. Imago with abdominal margin in male plaited, but not reflexed; body weak; antennæ long; wings much dilated, often tailed.

  d. Ulysses-group.

  e. Peranthus-group. ⎫ Protenor - group (Indian) is
  f. Memnon-group. ⎬ somewhat intermediate between these, and is nearest to the Nox-group.

  g. Helenus-group.
  h. Erectheus-group.
  i. Pammon-group.
  k. Demolion-group.

C. Larvæ subcylindrical, variously coloured. Imago with abdominal margin in male plaited, but not reflexed; body weak; antennæ short, with a thick curved club; wings entire.

l. Erithonius-group. Sexes alike, larva and pupa something like those of P. Demolion.

m. Paradoxa-group. Sexes different.

n. Dissimilis - group. Sexes alike; larva bright - coloured; pupa straight, cylindric.

D. Larvæ elongate, attenuate behind, and often bifid, with lateral and oblique pale stripes, green. Imago with the abdominal margin in male reflexed, woolly or hairy within; anal valves small, hairy; antennæ short, stout; body stout.

o. Macareus-group. Hind wings entire.

p. Antiphates-group. Hind wings much tailed (swallow-tails).

q. Eurypylus-group. Hind wings elongate or tailed.

Genus LEPTOCIRCUS.

Making, in all, twenty distinct groups of Malayan Papilionidæ.

The first section of the genus Papilio (A) comprises insects which, though differing considerably in structure, having much general resemblance. They all have a weak, low flight, frequent the most luxuriant forest-districts, seem to love the shade, and are the objects of mimicry by other Papilios.

Section B consists of weak-bodied, large-winged insects, with an irregular wavering flight, and which, when resting on foliage, often expand the wings, which the species of the other sections rarely or never do. They are the most conspicuous and striking of eastern Butterflies.

Section C consists of much weaker and slower-flying insects, often resembling in their flight, as well as in their colours, species of Danaidæ.

Section D contains the strongest-bodied and most swift-flying of the genus. They love sunlight, and frequent the borders of streams and the edges of puddles, where they gather together in swarms consisting of several species, greedily sucking up the moisture, and, when disturbed, circling round in the air, or flying high and with great strength and rapidity.

Geographical Distribution.--One hundred and thirty species of Malayan Papilionidæ are now known within the district extending from the Malay peninsula, on the north-west, to Woodlark Island, near New Guinea, on the south-east.

The exceeding richness of the Malayan region in these fine insects is seen by comparing the number of species found in the different tropical regions of the earth. From all Africa only 33 species of Papilio are known; but as several are still undescribed in collections, we may raise their number to about 40. In all tropical Asia there are at present described only 65 species, and I have seen in collections but two or three which have not yet been named. In South America, south of Panama, there are 150 species, or about one-seventh more than are yet known from the Malayan region; but the area of the two countries is very different; for while South America (even excluding Patagonia) contains 5,000,000 square miles, a line encircling the whole of the Malayan islands would only include an area of 2,700,000 square miles, of which the land-area would be about 1,000,000 square miles. This superior richness is partly real and partly apparent. The breaking up of a district into small isolated portions, as in an archipelago, seems highly favourable to the segregation and perpetuation of local peculiarities in certain groups; so that a species which on a continent might have a wide range, and whose local forms, if any, would be so connected together that it would be impossible to separate them, may become by isolation reduced to a number of such clearly defined and constant forms that we are obliged to count them as species. From this point of view, therefore, the greater proportionate number of Malayan species may be considered as apparent only. Its true superiority is shown, on the other hand, by the possession of three genera and twenty groups of Papilionidæ against a single genus and eight groups in South America, and also by the much greater average size of the Malayan species. In most other families, however, the reverse is the case, the South American Nymphalidæ, Satyridæ, and Erycinidæ far surpassing those of the East in number, variety, and beauty.

The following list, exhibiting the range and distribution of each group, will enable us to study more easily their internal and external relations.

Range of the Groups of Malayan Papilionidæ.

ORNITHOPTERA.

1. Priamus-group. Moluccas to Woodlark Island .... 5 species.

2. Pompeus-group. Himalayas to New Guinea, (Celebes, maximum) .... 11 species.

3. Brookeana-group. Sumatra and Borneo .... 1 species.

PAPILIO.

4. Nox-group. North India, Java, and Philippines .... 5 species.

5. Coon-group. North India to Java .... 2 species.

6. Polydorus-group. India to New Guinea and Pacific... 7 species.

7. Ulysses-group. Celebes to New Caledonia .... 4 species.

8. Peranthus-group. India to Timor and Moluccas (India, maximum) .... 9 species.

9. Memnon-group. India to Timor and Moluccas (Java, maximum) .... 10 species.

10. Helenus-group. Africa and India to New Guinea .. 11 species.

11. Pammon-group. India to Pacific and Australia .... 9 species.

12. Erectheus-group. Celebes to Australia .... 8 species.

13. Demolion-group. India to Celebes .... 2 species.

14. Erithonius-group. Africa, India, Australia .... 1 species.

15. Paradoxa-group. India to Java (Borneo, maximum) 5 species.

16. Dissimilis-group. India to Timor (India, maximum) 2 species.

17. Macareus-group. India to New Guinea .... 10 species.

18. Antiphates-group. Widely distributed .... 8 species.

19. Eurypylus-group. India to Australia..................... 15 species.

LEPTOCIRCUS.

20. Leptocircus-group. India to Celebes .... 4 species.

This Table shows the great affinity of the Malayan with the Indian Papilionidæ, only three out of the nineteen groups ranging beyond, into Africa, Europe, or America. The limitation of groups to the Indo-Malayan or Austro-Malayan divisions of the archipelago, which is so well marked in the higher animals, is much less conspicuous in insects, but is shown in some degree by the Papilionidæ. The following groups are either almost or entirely restricted to one portion of the archipelago:--

Indo-Malayan Region.	*Austro-Malayan Region.*
Nox-group.	Priamus-group.
Coon-group.	Ulysses-group.
Macareus-group (nearly).	Erechtheus-group.
Paradoxa-group.	
Dissimilis-group (nearly).	
Brookeanus-group.	
LEPTOCIRCUS (genus).	

The remaining groups, which range over the whole archipelago, are, in many cases, insects of very powerful flight, or they frequent open places and the sea-beach, and are thus more likely to get blown from island to island. The fact that three such characteristic groups as those of Priamus, Ulysses, and Erechtheus are strictly limited to the Australian region of the archipelago, while five other groups are with equal strictness confined to the Indian region, is a strong corroboration of that division which has been founded almost entirely on the distribution of Mammalia and Birds.

If the various Malayan islands have undergone recent changes of level, and if any of them have been more closely united within the period of existing species than they are now, we may expect to find indications of such changes in community of species between islands now widely separated; while those islands which have long remained isolated would have had time to acquire peculiar forms by a slow and natural process of modification.

An examination of the relations of the species of the adjacent islands, will thus enable us to correct opinions formed from a mere consideration of their relative positions. For example, looking at a map of the archipelago, it is almost impossible to avoid the idea that Java and Sumatra have been recently united; their present proximity is so great, and they have such an obvious resemblance in their volcanic structure. Yet there can be little doubt that this opinion is erroneous, and that Sumatra has had a more recent and more intimate connexion with Borneo than it has had with Java. This is strikingly shown by the mammals of these islands--very few of the species of Java and Sumatra being identical, while a considerable number are common to Sumatra and Borneo. The birds show a somewhat similar relationship; and we shall find that the distribution of the Papilionidæ tells exactly the same tale. Thus:--

Sumatra has...	21 species	}	20 sp. common to both islands;
Borneo ,, ...	30 ,,		
Sumatra ,, ...	21 ,,	}	11 sp. common to both islands;
Java ,, ...	28 ,,		
Borneo ,, ...	30 ,,	}	20 sp. common to both islands;
Java ,, ...	28 ,,		

showing that both Sumatra and Java have a much closer relationship to Borneo than they have to each other--a most singular and interesting result, when we consider the wide separation of Borneo from them both, and its very different structure. The evidence furnished by a single group of insects would have had but little weight on a point of such magnitude if standing alone; but coming as it does to confirm deductions drawn from whole classes of the higher animals, it must be admitted to have considerable value.

We may determine in a similar manner the relations of the different Papuan Islands to New Guinea. Of thirteen species of Papilionidæ obtained in the Aru Islands, six were also found in New Guinea, and seven not. Of nine species obtained at Waigiou, six were New Guinea, and three not. The five species found at Mysol were all New Guinea species. Mysol, therefore, has closer relations to New Guinea than the other islands; and this is corroborated by the distribution of the birds, of which I will only now give one instance. The Paradise Bird found in Mysol is the common New Guinea species, while the Aru Islands and Waigiou have each a species peculiar to themselves.

The large island of Borneo, which contains more species of Papilionidæ than any other in the archipelago, has nevertheless only three peculiar to itself; and it is quite possible, and even probable, that one of these may be found in Sumatra or Java. The last-named island has also three species peculiar to it; Sumatra has not one, and the peninsula of Malacca only two. The identity of species is even greater than in birds or in most other groups of insects, and points very strongly to a recent connexion of the whole with each other and the continent.

Remarkable Peculiarities of the Island of Celebes.

If we now pass to the next island (Celebes), separated from those last mentioned by a strait not wider than that which divides them from each other, we have a striking contrast; for with a total number of species less than either Borneo or Java, no fewer than eighteen are absolutely restricted to it. Further east, the large islands of Ceram and New Guinea have only three species peculiar to each, and Timor has five. We shall have to look, not to single islands, but to whole groups, in order to obtain an amount of individuality comparable with that of Celebes. For example, the extensive group comprising the large islands of Java, Borneo, and Sumatra, with the peninsula of Malacca, possessing altogether 48 species, has about 24, or just half, peculiar to it; the numerous group of the Philippines possess 22 species, of which 17 are peculiar; the seven chief islands of the Moluccas have 27, of which 12 are peculiar; and the whole of the Papuan Islands, with an equal number of species, have 17 peculiar. Comparable with the most isolated of these groups is Celebes, with its 24 species, of which the large proportion of 18 are peculiar. We see, therefore, that the opinion I have elsewhere expressed, of the high degree of isolation and the remarkable distinctive features of this interesting island, is fully borne out by the examination of this conspicuous family of insects. A single straggling island with a few small satellites, it is zoologically of equal importance with extensive groups of islands many times as large as itself; and standing in the very centre of the archipelago, surrounded on every side with islets connecting it with the larger groups, and which seem to afford the greatest facilities for the migration and intercommunication of their respective productions, it yet stands out conspicuous with a character of its own in every department of nature, and presents peculiarities which are, I believe, without a parallel in any similar locality on the globe.

Briefly to summarize these peculiarities, Celebes possesses three genera of mammals (out of the very small number which inhabit it) which are of singular and isolated forms, viz., Cynopithecus, a tailless Ape allied to the Baboons; Anoa, a straight-horned Antelope of obscure affinities, but quite unlike anything else in the whole archipelago or in India; and Babirusa, an altogether abnormal wild Pig. With a rather limited bird population, Celebes has an immense preponderance of species confined to it, and has also six remarkable genera (Meropogon, Ceycopsis, Streptocitta, Enodes, Scissirostrum, and Megacephalon) entirely restricted to its narrow limits, as well as two others (Prioniturus and Basilornis) which only range to a single island beyond it.

Mr. Smith's elaborate tables of the distribution of Malayan Hymenoptera (see "Proc. Linn. Soc." Zool. vol. vii.) show that out of the large number of 301 species collected in Celebes, 190 (or nearly two-thirds) are absolutely restricted to it, although Borneo on one side, and the various islands of the Moluccas on the other, were equally well explored by me; and no less than twelve of the genera are not found in any other island of the archipelago. I have shown in the present essay that, in the Papilionidæ, it has far more species of its own than any other island, and a greater proportion of peculiar species than many of the large groups of islands in the archipelago--and that it gives to a large number of the species and varieties which inhabit it, 1st, an increase of size, and, 2nd, a peculiar modification in the form of the wings, which stamp upon the most dissimilar insects a mark distinctive of their common birth-place.

What, I would ask, are we to do with phenomena such as these? Are we to rest content with that very simple, but at the same time very unsatisfying explanation, that all these insects and other animals were created exactly *as* they are, and originally placed exactly *where* they are, by the inscrutable will of their Creator, and that we have nothing to do but to register the facts and wonder? Was this single island selected for a fantastic display of creative power, merely to excite a childlike and unreasoning admiration? Is all this appearance of gradual modification by the action of natural causes--a modification the successive steps of which we can almost trace--all delusive? Is this harmony between the most diverse groups, all presenting analogous phenomena, and indicating a dependence upon physical changes of which we have independent evidence, all false testimony? If I could think so, the study of nature would have lost for me its greatest charm. I should feel as would the geologist, if you could convince him that his interpretation of the earth's past history was all a delusion--that strata were never formed in the primeval ocean, and that the fossils he so carefully collects and studies are no true record of a former living world, but were all created just as they now are, and in the rocks where he now finds them.

I must here express my own belief that none of these phenomena, however apparently isolated or insignificant, can ever stand alone--that not the wing of a butterfly can change in form or vary in colour, except in harmony with, and as a part of the grand march of nature. I believe, therefore, that all the curious phenomena I have just recapitulated, are immediately dependent on the last series of changes, organic and inorganic, in these regions; and as the phenomena presented by the island of Celebes differ from those of all the surrounding islands, it can, I conceive, only be because the past history of Celebes has been, to some extent, unique and different from theirs. We must have much more evidence to determine exactly in what that difference has consisted. At present, I only see my way clear to one deduction, viz., that Celebes represents one of the oldest parts of the archipelago; that it has been formerly more completely isolated both from India and from Australia than it is now, and that amid all the mutations it has undergone, a relic or substratum of the fauna and flora of some more ancient land has been here preserved to us.

It is only since my return home, and since I have

been able to compare the productions of Celebes side by side with those of the surrounding islands, that I have been fully impressed with their peculiarity, and the great interest that attaches to them. The plants and the reptiles are still almost unknown; and it is to be hoped that some enterprising naturalist may soon devote himself to their study. The geology of the country would also be well worth exploring, and its newer fossils would be of especial interest as elucidating the changes which have led to its present anomalous condition. This island stands, as it were, upon the boundary-line between two worlds. On one side is that ancient Australian fauna, which preserves to the present day the facies of an early geological epoch; on the other is the rich and varied fauna of Asia, which seems to contain, in every class and order, the most perfect and highly organised animals. Celebes has relations to both, yet strictly belongs to neither: it possesses characteristics which are altogether its own; and I am convinced that no single island upon the globe would so well repay a careful and detailed research into its past and present history.

Concluding Remarks.

In writing this essay it has been my object to show how much may, under favourable circumstances, be learnt by the study of what may be termed the external physiology of a small group of animals, inhabiting a limited district. This branch of natural history had received little attention till Mr. Darwin showed how important an adjunct it may become towards a true interpretation of the history of organized beings, and attracted towards it some small share of that research which had before been almost exclusively devoted to internal structure and physiology. The nature of species, the laws of variation, the mysterious influence of locality on both form and colour, the phenomena of dimorphism and of mimicry, the modifying influence of sex, the general laws of geographical distribution, and the interpretation of past changes of the earth's surface, have all been more or less fully illustrated by the very limited group of the Malayan Papilionidæ; while, at the same time, the deductions drawn therefrom have been shown to be supported by analogous facts, occurring in other and often widely-separated groups of animals.

Note Appearing in the Original Work

1. W. O. Hewitson, Esq., of Oatlands, Walton-on-Thames, author of "Exotic Butterflies" and several other works, illustrated by exquisite coloured figures drawn by himself; and owner of the finest collection of Butterflies in the world.

www.ingramcontent.com/pod-product-compliance
Lightning Source LLC
Chambersburg PA
CBHW021626270326
41931CB00008B/883

* 9 7 8 1 4 7 3 3 2 9 8 1 2 *